~ World Winds

Written by Heather Fleming ~ Illustrated by Marissa Abrego

© 2025 Park Publishing

ISBN 979-8-9882902-1-6

All rights reserved. No part of this may be reproduced, stored in a retrieval system, or transmitted in any form or by any means - electronic, mechanical, photocopy, recording or any other - except for brief quotations in printed reviews, without the prior permission of the publisher. © 2025 Park Publishing

What is wind?

How do you describe something you can't see?

We see the effects of wind all around us. For example, you <u>see</u> flags blowing in the wind. You <u>feel</u> hot wind brush past your cheeks in the summer or cold wind bite at your nose in the winter. We <u>hear</u> trees rustle when the wind blows past their branches, but what <u>causes</u> wind?

In science, heat rises and cold sinks. This is the same with air. Warm air wants to rise high into the sky, while cold air wants to sink down close to the ground.

If the ground is warm, it makes the air above it warm also. When cool wind blows in, this causes the two types of air to be thrown off balance. Warm air on the ground wants to change places with the cold air above it. This pushes the two types of air past each other, which causes wind.

Whenever both the ground air temperature and the sky air temperature are the same, there is no wind.

Throughout this book we use poetry from different countries' specific regions to better understand the effects of wind across our world.

Kara-Kaze

Mount Haruna halts
Snowy winds can't pass it's peak
Dry winter blows through

Bize

Driven by a cold front a bize will bring,
Crisp dry air from a northern direction.
Typically lasting three days in the spring,
Covering land it makes a connection.
The Jura and Alps form a cross-section,
Where southern winds blow in heavy weather.
Whirling icy winds leave no protection,
As storms from the mountains come together.

Buran

<u>B</u>lustery cold winds kick up snow in winter
<u>U</u>nleashing blizzard-like conditions down below
<u>R</u>obust wind swirls snow until you cannot see
<u>A</u> bitter chill pierces through winter gear
<u>N</u>ortheasterly wind brings ice and snow

Brickfielder

From the outback heated and robust,
Willys form in red clouds of dust.
Dry winds stir up a whirl
causing a rowdy swirl
which makes holding your hat a must!

KasKazi-Kusi

Pairs of winds blowing from opposite sides,
Pushing and shifting as they take their hold.
We count on changes their weather provides,
For one blows hot and the other blows cold.
Mighty north wind, your name is KasKazi,
Hot and dry, from Arabia streaming.
Sweltering heat makes the land so hazy,
Of cooler days, our minds are daydreaming.
Soothing Southern wind, your name is Kusi,
From the Ocean of India you flow.
Cooler breezes case our lives as you see,
The rain you provide prompts our crops to grow.
These two winds give us plenty of reasons,
To dance and sing in these changing seasons.

PAMPERO

Rushing over the Andes
Across the flat plains,
Blows the cold, gusty, dry wind
Bringing with it rains.
Mixing with wet air,
A fierce squall line develops
Storming everywhere.

Monsoon

Looming above the land, Himalayans stand
Blocking monsoon wind, causing the rains to end.
Advancing with all might, wind loses the fight
And in defeat drops showers until it stops.
Now dry air can rise and limp to northern side.
Gathering up speed, the wind grows swift indeed.
Keeping the north dry, even though it does try
To bring with it rain, yet it never can gain
Strength to pick up showers without much power.

Loo

Oh Loo, what plans have you begun today?
We are so tired and quite done today.
The Andhi this spring was brutal on us,
That fierce storm we couldn't outrun today.
A wall of dust blew all over our land,
Filling the sky, it blocked the sun today.
Now it is summer, we hope for the best,
A break from the heat, we want fun today.
Oh Loo, heat from you causes stress on us,
Heatstroke and headaches, you spare none today.
The only relief comes after you're through.
Monsoon rains have you on the run today.

Sirocco

Dry
Heat
Swirling
Hot winds blow
Creating sandstorms
Constantly shifting sands build dunes

Katabatic

Elevated ice cools surrounding air,
Dense winds blow across plains, leaving them bare.
Interior plateaus slow down the race,
But gravity pulls as wind picks up the pace.
Steep coastal slopes are where the winds descend,
Hurtling downward, the blast has no end.
Emperor penguins huddle up in mass,
As powerfully strong winds push to pass.
Blowing snow away to show ice the most,
Katabatic winds finally reach the coast.

Chinook

West
cool, wet
climbing, drying, blowing
Rocky Mountains snow eater
descending, heating, melting
warm, dry
East

LOCATION	NAME OF WIND	PRONUNCIATION	TYPE OF POEM
Himalayan Mountains	Monsoon	mon-soon	Masnavi
Kenya	KasKazi - Kusi	kas-kazi - ku-zee	Sonnet
Rocky Mountains	Chinook	chuh-nuk	Diamante
Antarctica	Katabatic	kat-uh-bad-ik	Couplet
Sahara Desert	Sirocco	sir-ah-co	Fibonacci
India	Loo	loo	Ghazal
Russia	Buran	boo-rahn	Acrostic
Australia	Brickfielder	brick-feel-duhr	Limerick
Japan	Kara Kaze	care-u-kah-zee	Haiku
Switzerland	Bize	beez	Huitain
Argentina	Pampero	pam-pair-oh	Sequidilla

HAIKU
(hi-koo)

- Form of Japanese poetry
- Theme is usually about nature
- Contains a total of three lines
  - First line = 5 syllables
  - Second line = 7 syllables
  - Third line = 5 syllables

Huitain
(we-tin)

- Form of French poetry
- Contains 8 lines
- 8 – 10 syllables per line
- Pattern of AB/AB/BC/BC

Acrostic
(uh-kros-tik)

- Form of ancient Greek and Latin poetry
- A message is formed down the side of the poem with the first letter of each verse
- Poem does not need to rhyme

Limerick
(lim-er-ik)

- Form of British poetry
- Theme usually is humorous
- Contains 5 lines
- Lines 1 and 2 = 8 or 9 syllables
- Lines 3 and 4 = 5 or 6 syllables
- Line 5 = 8 or 9 syllables
- Pattern of AA/BB/A

Sonnet
(sau-nit)

- **Form of Mediterranean poetry**
- **Halfway through a sonnet, the tone or mood changes**
- **Contains 14 lines**
- **Each line contains 10 syllables**
- **Pattern of a common sonnet is A/B/A/B/C/D/C/D/E/F/E/F/G/G**

Sequidilla
(se-gee-dee-ya)

- Form of Spanish poetry
- Contains 7 lines
- Line 1, 3, 5 = 7 syllables
- Line 2, 4, 5, 7 = 5 syllables
- Lines 2 and 4 rhyme
- Lines 5 and 7 rhyme
- There is a change of direction between lines 4 and 5

Ghazal
(guz-all)

- Form of Indian poetry
- 5 to 15 couplets - (two lines)
- The first two lines must end in the same word
- Every couplet after that must also end in that word.
- The second word before the last word in each couplet must rhyme.

Masnavi
(Mas-nah-vee)

- Form of Persian poetry
- Based on a story or subject
- No limit on the amount of lines
- Lines form couplets – two rhyming lines together
- Rhyming happens two times in each line – called a 'half-line rhyme'
- Each line contains 11 syllables
- Pattern of AA/BB/CC/DD/EE

Fibonacci
(fee-bo-na-chee)

- Form of Italian poetry
- Each number is made by adding up the two previous numbers...

$$0 + 1 = 1$$

$$1 + 1 = 2$$

$$1 + 2 = 3$$

$$2 + 3 = 5$$

$$3 + 5 = 8$$

This pattern continues on and on...

- Syllables are matched to the Fibonacci sequence:

0,1,1,2,3,5,8,13,21,

- No limit to the number of lines

Couplet
(kuhp-lit)

- Form of poetry from almost every country
- Two lines that usually rhyme
- Can group several couplets in one poem
- Pattern of A/A, B/B

Diamante
(dee-uh-mahn-tey)

- 7 lines
- Unrhymed shape poem
- Begins and ends with opposite nouns
- First line = 1 noun
- Second line = 2 adjectives
- Third line = 3 verbs
- Fourth line = 4 nouns (shift happens here)
- Fifth line = 3 verbs
- Sixth line = 2 adjectives
- Seventh line = 1 noun

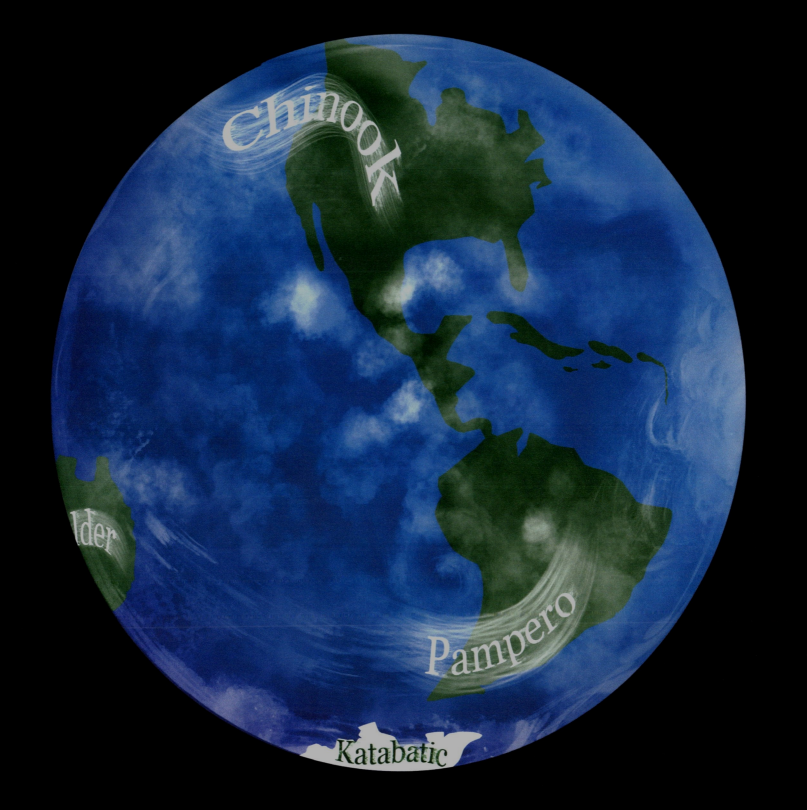

www.ingramcontent.com/pod-product-compliance
Lightning Source LLC
LaVergne TN
LVHW071102060425
807841LV00016B/30